STEAM-Powered Careers

OCCUPATIONAL THERAPY

by **Jasmin Sanchez**, featured scientist

illustrated by **Michelle Laurentia Agatha**

Room to Read

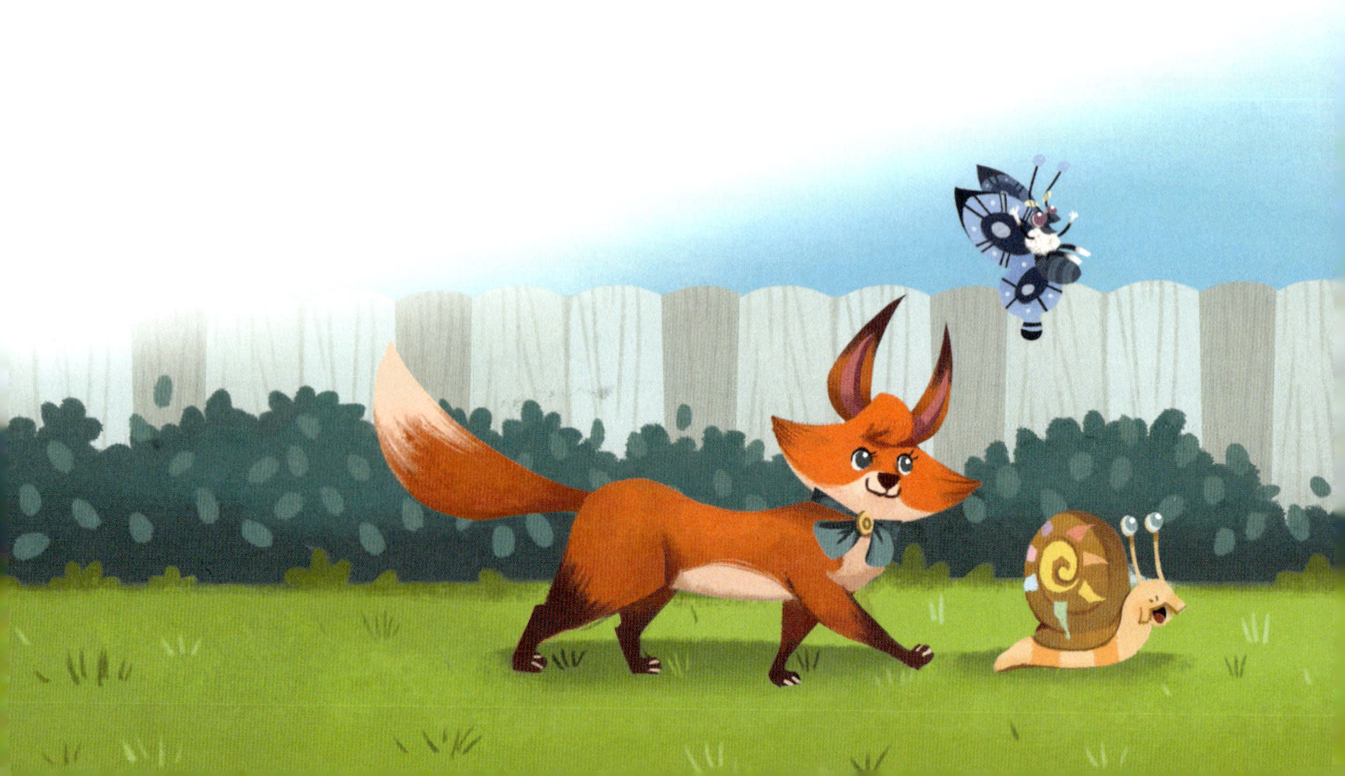

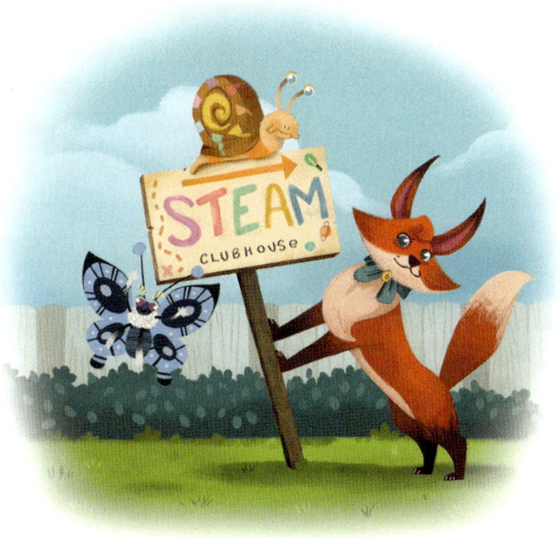
STEAM
CLUBHOUSE

*To my brother, Jessie. This book was made with you in mind.
You have taught me so much over the years, and you are the reason
I pursue occupational therapy every day.
A special thank-you to my mom, Dr. Dieuwertje DJ Kast, Dr. Payne, Jason, Carlos,
and all my friends who've read my book cover to cover! —J.S.*

*For my kind and generous little sister,
your presence is always a blessing for others —M.L.A.*

Room to Read would like to thank Tatcha™ for their generous
support of the STEAM-Powered Careers collection.

TATCHA

Copyright 2022 Room to Read. All rights reserved.

Written by Jasmin Sanchez
Featured scientist: Jasmin Sanchez
Illustrated by Michelle Laurentia Agatha
Edited by Jocelyn Argueta
Photo research by Kris Durán
Series art direction and design by Christy Hale
Series edited by Carol Burrell, Jamie Leigh Real, Jocelyn Argueta, and Deborah Davis
Copyedited by: Debra Deford-Minerva and Danielle Sunshine

ISBN 978-1-63845-064-1

Manufactured in Canada.

MIX
Paper | Supporting responsible forestry
FSC® C011825

10 9 8 7 6 5 4 3 2

Room to Read
465 California Street #1000
San Francisco, California 94104
roomtoread.org

Room to Read

World change starts with educated children.©

Contents

Explore Occupational Therapy with Mia and Sunny	6
What Is Occupational Therapy?	22
Meet the Scientist	24
Learn More about Occupational Therapy	30
Word List	34

Mia rises from a beanbag chair in the clubhouse, yawns, and stretches. She turns to wake up Sunny from his nap.

Sunny gets up quickly. "That was a great nap. Now that we are rested, let's get ready for an adventure—but first we have to brush our teeth."

Mia and Sunny go over to the sink.

"Remember when you needed help brushing your teeth?" Sunny asks.

"Yes," Mia says, "but I'm finally able to do it on my own!"

Mia notices that Sunny is not able to pick up his toothbrush. "It seems like you still need a little help there."

Sunny frowns. "Yes."

"I have an idea!" Mia says. "We can figure out a different way for you to brush your teeth."

Sunny smiles and nods.

"After all," Mia explains, "brushing your teeth is super important. It's considered one of the **Activities of Daily Living (ADLs)**, actions we have to do every day to live and be healthy. Last week we had a career fair at school, and I met a very interesting person who is an occupational therapist. She told me that there isn't one perfect way to do anything, because there are a million ways to do everything."

> Occupational therapists help people do their ADLs, such as eating, taking a shower, and moving from one place to another, to prepare them to live on their own.

STEAM-Powered Careers

Mia takes Sunny's toothbrush and sticks a piece of strong sticky tape to the back.

"What if we tape your toothbrush to the wall?" she asks Sunny.

"That would be great," Sunny replies. "Then I can brush my teeth without asking you to hold it."

Mia sticks the toothbrush to the wall and Sunny begins to brush his teeth.

"This is terrific!" he says. "Now even if you're not here, I can brush my teeth on my own."

Mia quickly unzips her backpack and takes out her super-soft sweater. She puts it on and buttons it, one button after another.

Sunny watches Mia with amazement. "Wow! When did you learn to unzip your backpack and button your sweater so fast?"

"I learned by practicing my **fine motor skills**," Mia says. "The occupational therapist taught me that we practice our fine motor skills in school all the time, like when we use scissors to cut paper, when we hold our pencils with our fingers, and even when we play with LEGO bricks or puzzles. By using our fingers, we strengthen our hand muscles. And the more we practice, the better we get."

> We develop fine motor skills when we practice the small and very important movements of our hands.

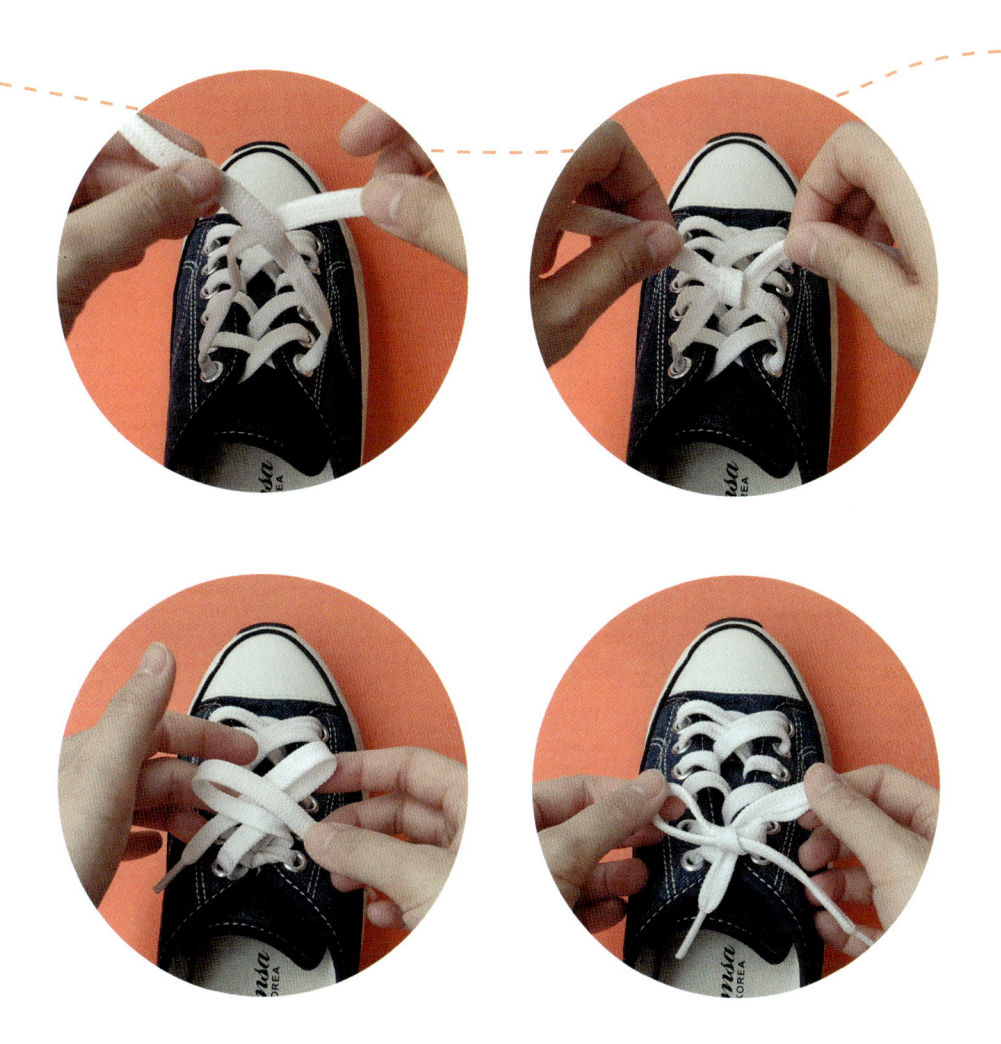

Mia looks down at her shoe and realizes the lace is untied.

"Oh, no," Sunny says. "There isn't an adult around to help."

"It's OK. I can tie my own shoes now. First, I cross the laces and make an X. I fold one lace under the lower part of the X and pull the lace out. Then, I make two bunny ears—one with my right hand and one with my left hand. Finally, I cross the bunny ears, fold one under the other, and pull."

"Wow!" Sunny says. "Those fine motor skills are coming in handy."

Mia giggles. "Yes! Thank you, Sunny. I've been practicing them!"

Modified items have been improved to help people with different ability levels do activities they like, or activities they need to do.

"Hmm," Sunny says, "all this learning is making me hungry."

"Me too." Mia unzips her lunch bag and takes out two pieces of bread. She doesn't have a fork to spread her peanut butter, so she and Sunny search the clubhouse for utensils.

Sunny finds a fork that looks different. "What's this?"

"It's a **modified** fork," Mia explains. "It's a fork that everyone can use. Sometimes people can't move their wrists or fingers to use a traditional fork. A person may have a **disability** or a hand injury that doesn't let them use their fine motor skills."

"Wow, that fork is amazing," Sunny says. "So when we added tape to my toothbrush earlier, did we modify it so I could use it?"

"Yes!" Mia says. "Aha! I can use the modified fork to spread my peanut butter, since anyone can use it!"

After spreading the peanut butter, she opens a container of strawberry and banana slices. She uses the fork to pick them out and place them on the bread. She also takes out a bowl of leaves and grass for Sunny. Together, they enjoy a healthy meal.

After lunch, Mia puts on a white lab coat that has "Future Occupational Therapist Mia" stitched on it. They leave the clubhouse, ready for adventure.

"Wow, look at you working your fine motor skills by walking down the ramp," Sunny says.

"Not quite!" Mia says. "When I move my leg muscles, I am actually using my **gross motor skills**."

"Are gross motor skills different from fine motor skills?" Sunny asks.

"Yes, fine motor skills are for small muscles, like in fingers and wrists, while gross motor skills are for big muscles, like in legs and arms."

"So you use fine motor skills when you grab your water bottle, and gross motor skills when you skip or run during recess?" Sunny asks.

"Exactly," Mia says. "And I know the perfect way I can practice my gross motor skills."

She runs to the bike station, and Sunny follows.

> You use your gross motor skills when doing jumping jacks.

Occupational Therapy 15

Before they get on Mia's bike, Sunny notices that it has three wheels, a back support, and pedal straps. The other bike on the rack has only two wheels, and it doesn't have a back support or pedal straps.

Sunny asks, "Is your bike modified?"

"Yes," Mia says. "It's easier for me to ride a tricycle."

Sunny smiles. "I love your tricycle. It will help us get places quickly and safely."

They put on their helmets, and Mia starts to pedal her tricycle.

"This is so much fun!" Sunny cheers.

"And riding a bike is also good for our bodies," Mia says. "It helps our bodies stay healthy and makes our muscles strong. We are practicing our fine motor skills when we grip the handlebars, and—"

Sunny interrupts excitedly. "You're practicing your gross motor skills when you use your leg muscles to push the pedals!"

"Yes!"

As they make their way back toward the clubhouse, Mia and Sunny approach a high curb.

"Oh, no," Mia says. "This curb is so high, we have to get off the bike to make it to the other side of the street."

They walk across the intersection and then get back on the bike.

"Hmm, that curb is not modified," Sunny says. "Not everyone can use it to cross to the other side."

"You're totally right and thinking like an occupational therapist," Mia says. "To fix this, we should make a report to the city's department of **public works**, so they can cut the curb and make it accessible for everyone."

Public works is a part of the government that helps the community fix roads and buildings.

Occupational Therapy 17

Once they reach the clubhouse, Mia and Sunny lock the bike on the rack there. They race up the ramp, and Sunny wins.

"That was so fun. I love ramps!" Sunny says.

"They can be fun, but the main job of a ramp is to keep people safe," Mia says.

"What do you mean by safe?"

"Ramps are made for people who can't use stairs because of a disability. Their illness or injury might stop them from using the stairs. They can use ramps with a safety railing instead. Not everyone can use the stairs, but everyone can use ramps that are **universally designed**."

"What does universally designed mean?" Sunny asks.

"Universal design is when something is modified so that everyone can use it. Another example is the doors at grocery stores that automatically open. That universal design helps people who have disabilities with their fine motor skills and can't use a doorknob or push the door open."

"Wow," Sunny says. "Universal design is so cool! I didn't know ramps were so helpful."

Universal design is a way of creating items, places, and environments so that anyone can use them without the help of others.

Occupational Therapy

Sunny and Mia go into the clubhouse and greet their friends. Sunny explains what gross motor skills, fine motor skills, modifications, and universal design are. Everyone is happy and excited to learn about what occupational therapists do.

Mia watches Sunny as he helps Jae tie his shoe and shows everyone the tricycle. Mia takes a label and writes on it, "Future Occupational Therapist Sunny."

Mia hands the label to her friend. "You have learned so much today, Sunny," she says, "and you're already helping others, just like an occupational therapist does."

The friends smile.

What Is Occupational Therapy?

Mia and Sunny have only skimmed across the surface of occupational therapy. Let's go over some terms that will be helpful when we tag along with **Jasmin Sanchez** for an inside look at how she helps people live their best lives and helps make the world a more inclusive place.

Occupational therapy is a field of study focused on helping people perform important and meaningful activities, even when they are faced with a challenge. These activities can include regular living routines such as brushing teeth, making pizza, and wiggling into a snuggly sweater. Meaningful activities can also include things that make us happy, such as playing a sport, talking with friends, drawing, and eating pizza.

Just like Mia, **Jasmin Sanchez** is studying to be an occupational therapist whose job is to help people overcome physical challenges caused by illness or injury. Let's ask Jasmin some questions before going with her to school and work!

Occupational Therapy

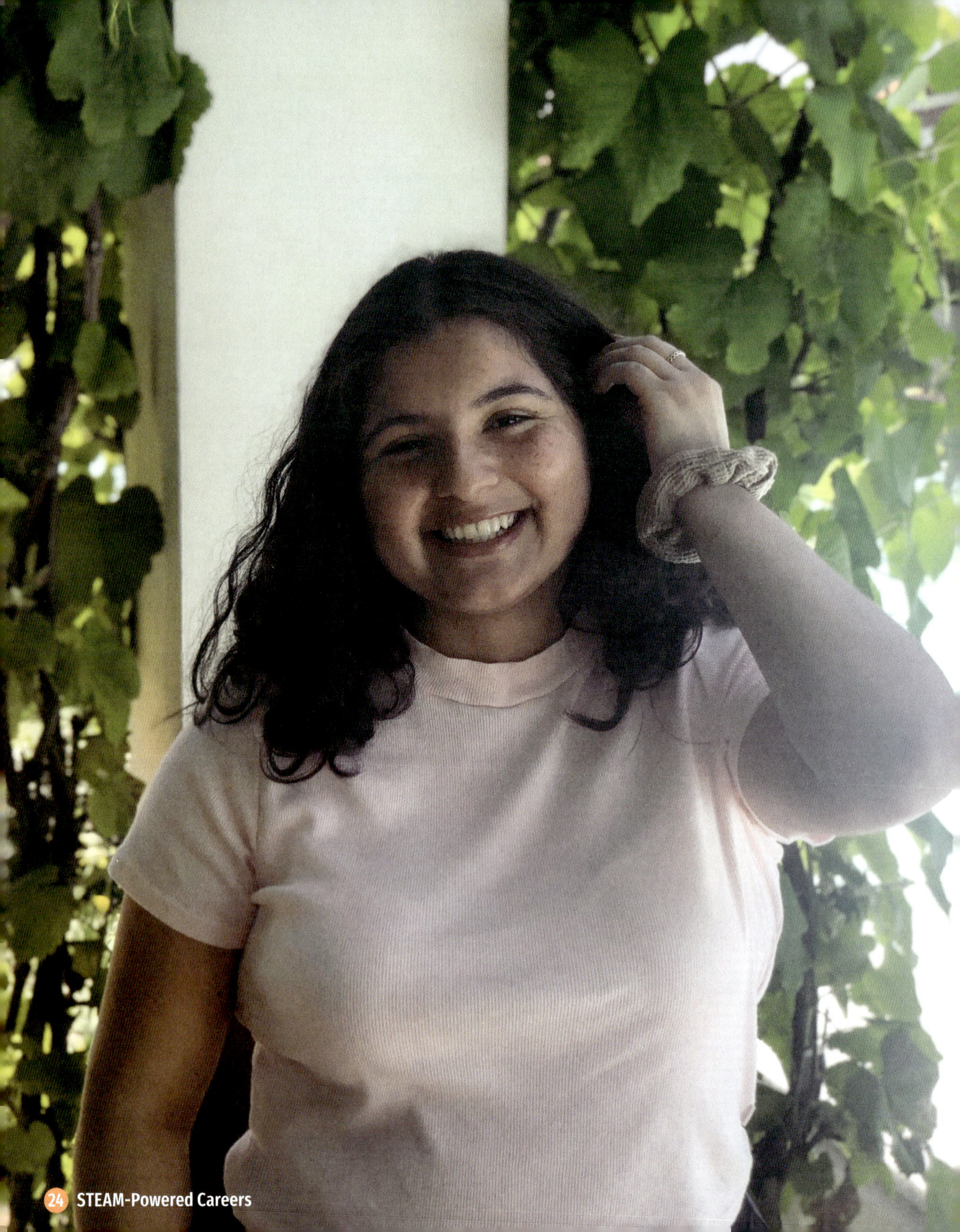

Meet the Scientist
Jasmin Sanchez

I am currently a senior at the University of Southern California, studying health and human science, with a minor in occupational science. I'll graduate in 2022.

Fun Fact #1: I loved doing science experiments as a kid, and I was part of a health academy in high school. I even competed in Future Health Professionals, formerly known as Health Occupation Students of America.

Fun Fact #2: I love art and creating things, so you'll always find me doing fun hands-on activities. I was in art classes and clubs in middle school and high school.

Fun Fact #3: I love the ocean and studying marine life.

What is your favorite thing about your field?

What is your least favorite thing about your field?

I love to help others and make them smile!

Dealing with medical insurance companies. It is incredibly hard, and not all people have access to an occupational therapist. I wish this was not the case.

I'll grab my backpack, and I'll show you what I do

A Day in the Life, Part 1

I stretch and meditate each day to ground myself and start the day feeling happy. Stretching helps me warm up the muscles I'll use throughout the day during fine motor activities and gross motor activities.

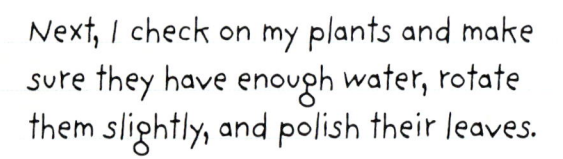

Next, I check on my plants and make sure they have enough water, rotate them slightly, and polish their leaves.

I brush my teeth and prepare a yummy and healthy breakfast.

Then I sit at my desk and study. I'm not an occupational therapist yet—I still have to continue with school and learn more. Every day, I learn something new about occupational therapy. One day soon I'll be able to go out into the world and help others.

I volunteer in a preschool with many different children. Together we practice our fine motor skills by cutting strands of construction paper and covering toys with the pieces of paper.

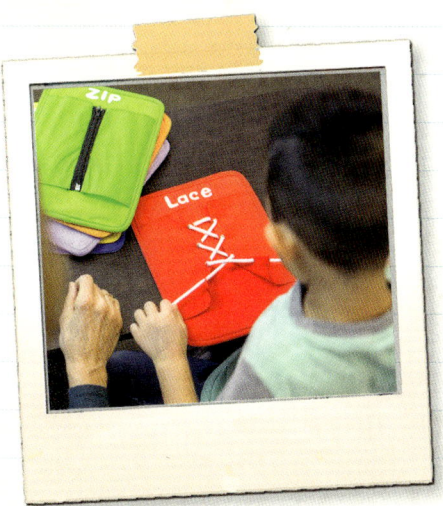

After the group work, the students each get to meet with an occupational therapist and discuss their Activities of Daily Living (ADLs). This helps the therapist work on a specific plan to help the student.

Occupational Therapy 27

A Day in the Life, Part 2

Occupational therapists work in a lot of different settings, but wherever they are, they generally follow these steps:

1. Meet with people who have questions or need help.
2. Listen to their challenges and make a plan for how to help them.
3. Teach the person the plan and adjust it together as needed.
4. Set the plan into motion and help people improve at their tasks every week.
5. If the plan doesn't work as expected, make adjustments and try again.

Some examples of what an occupational therapist can do are:

1. Throw bouncy balls to a person to help the child or adult with their fine motor skills.
2. Help a person feel more comfortable at work. They can add a backrest to a chair to improve posture, add a laptop stand to fix the eye level at their computer, or add a plant to their work space.

There are many different types of modified items in the world to help people do their Activities of Daily Living as well as other meaningful activities. For example:
- Modified utensils, such as spoons and forks
    - A strapped or side-handled fork or spoon can help people who are not able to move their fingers or wrists as much as others can.
- Balance beams
    - A balance beam can help people recovering from a back or leg injury.
- 4-Wheeled Posterior Walker and Gait Trainer
    - A special walker can help people who need extra walking support by supporting their body as they walk and move.
- Kaye Bolster Positioning Chairs for Trunk Alignment
    - These special chairs help people who need extra back and upper-body support sit up in classrooms and other places.

Occupational Therapy

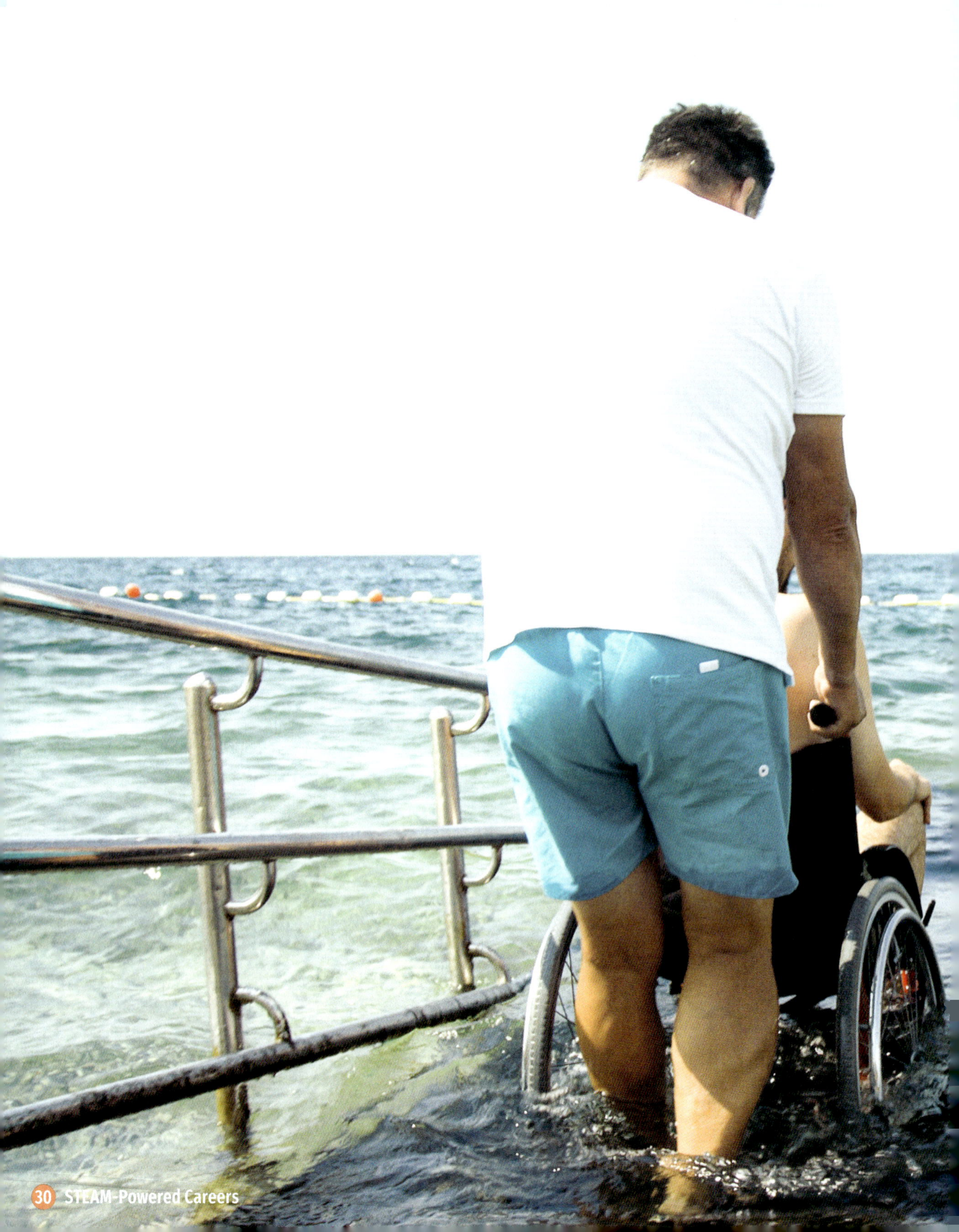

Occupational therapy needs people in a lot of related fields!

STEAM Careers in Occupational Therapy

Occupational therapists work together with many other professionals such as doctors, nurses, physical therapists, teachers, principals, speech pathologists, nutritionists, neurologists, and veterinarians.

An occupational therapist works either with one person at a time or with a group of people. They can work anywhere—in hospitals, schools, offices, and private homes—even outdoors.

The Future of Occupational Therapy

Universal Design: This is an exciting new field. Occupational scientists and therapists are working to make everything available to everyone. This means thinking carefully about the things we create and the places we spend time in, keeping in mind that people's needs and abilities will be different. If designers think about this when brainstorming and creating, then more people are empowered to use the item or space fully, without help.

Consider texting on your phone. It was originally designed for people who are Deaf and hard of hearing. And there are universally accessible playgrounds for children in Los Angeles. These parks allow for children to play together, no matter what their physical needs are.

Social Justice: This is another growing part of occupational therapy. Social justice is the idea that everyone should have the same access to money, opportunity, and safety. One way of working toward social justice is giving some communities resources and information to help them do what other communities are already doing to stay healthy.

Do You Want to Be an Occupational Therapist?

You've already done the most important part: caring about people!

Other things you can do now:
- Help other people.
- Modify things so that more people can use them.
- Support those around you in getting all the opportunities and materials they need to succeed.

After high school, you'll need an undergraduate degree (four years of college) and a master's degree or a doctorate degree (a PhD).

Word List

Activities of Daily Living (ADLs): actions we have to do every day to live and be healthy, such as eating, taking a shower, getting dressed, and using the bathroom

disability: an injury or illness that causes a person to need modifications in order to see, eat, smell, drink, walk, learn, or other activities

fine motor skills: actions done by small muscles in our hands; these develop when we practice delicate movements in our hands

gross motor skills: actions done using big muscles in our bodies; these develop when we practice large movements like sitting or walking

modified: when an item or space has been changed in a way that allows people to do activities they like or need to do

public works: a part of the government that focuses on helping the community fix roads and buildings

universally designed: items, products, or spaces that are made with the purpose of being accessible to as many people as possible

Occupational Therapy Resources

Going to an occupational therapist:
kidshealth.org/en/kids/occupational-therapist.html

How occupational therapy has helped:
youtu.be/Ud5Fp279g4Y

Universally accessible playgrounds in Los Angeles:
LAParks.org/uap

Universal design for learning explained with LEGO bricks:
youtu.be/KhMD2PDa6do

Acknowledgments
University of Southern California Joint Educational Project STEM Education Programs

Jasmin Sanchez is a senior at the University of Southern California (USC) majoring in health and human sciences and minoring in occupational science. She is a first-generation college student in her family who loves advocating for education, science, and physical accessibility. She is on her way to becoming an occupational therapist and aspires to help many people in the future with their occupational goals!

Michelle Laurentia Agatha was born in Jakarta, Indonesia. Ever since she was young, she has had a huge interest in cartoons and illustrated books. Michelle pursued her dream of becoming an illustrator by earning a Bachelor of Fine Arts degree from the Academy of Art University in San Francisco. Currently, Michelle is working as a children's book illustrator, concept artist, and UI/UX designer.

Explore the Complete

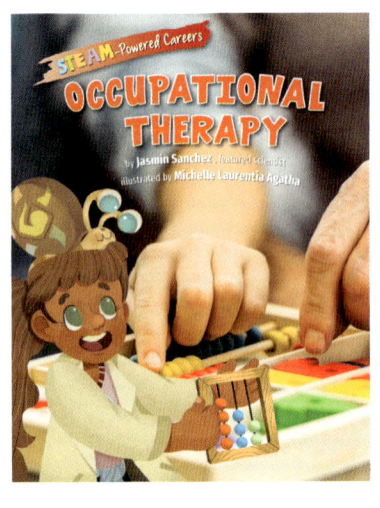

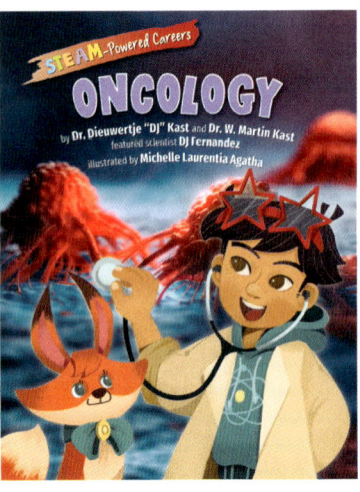

STEAM-Powered Careers: DATA SCIENCE
by Stacey Finley, featured scientist
illustrated by Michelle Laurentia Agatha

STEAM-Powered Careers: ENGINEERING
by Dr. Dijanna Figueroa, featured scientist Dr. Darin Gray
illustrated by Janet Pagliuca

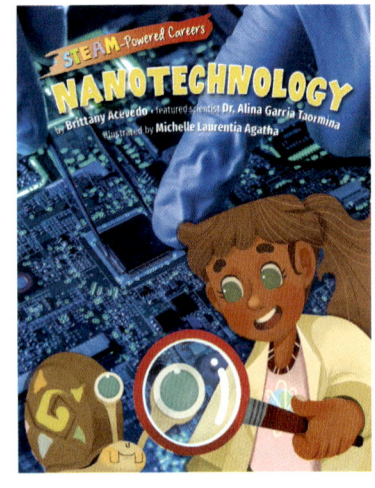

STEAM-Powered Careers: NANOTECHNOLOGY
by Brittany Acevedo, featured scientist Dr. Alina Garcia Taormina
illustrated by Michelle Laurentia Agatha

STEAM-Powered Careers: OCCUPATIONAL THERAPY
by Jasmin Sanchez, featured scientist
illustrated by Michelle Laurentia Agatha

STEAM-Powered Careers: ONCOLOGY
by Dr. Dieuwertje "DJ" Kast and Dr. W. Martin Kast
featured scientist DJ Fernandez
illustrated by Michelle Laurentia Agatha

STEAM-Powered Careers Series!

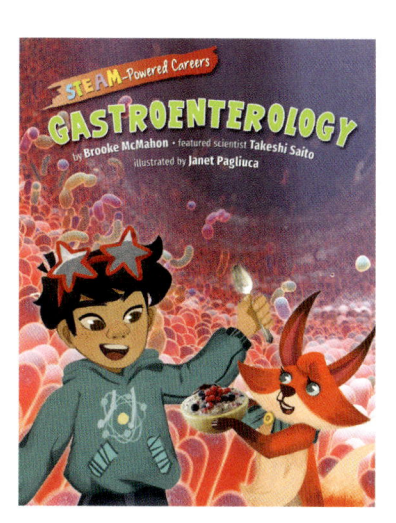

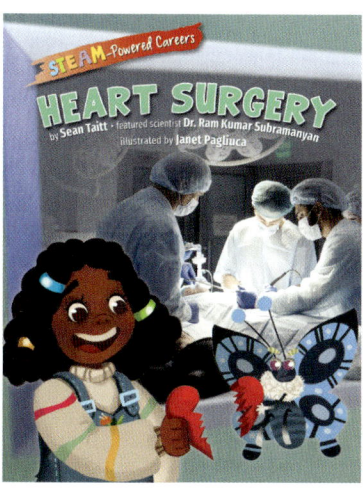

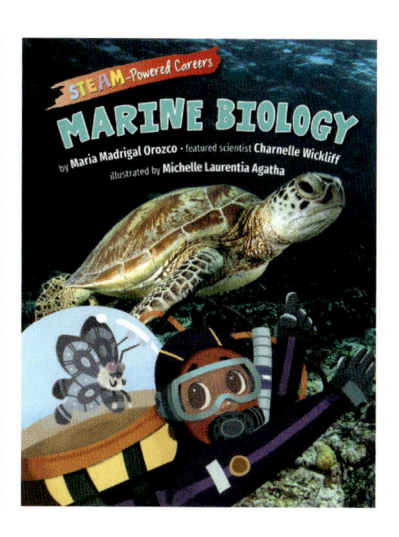

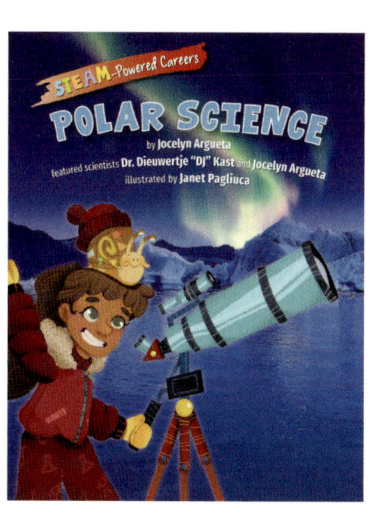

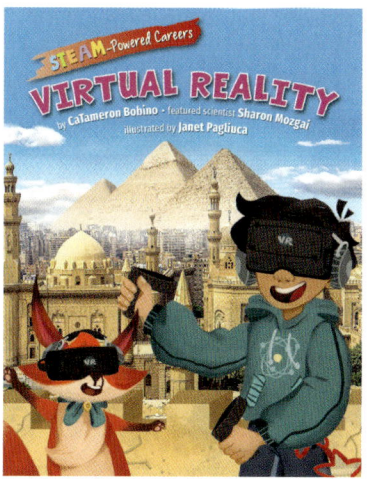

GASTROENTEROLOGY
by Brooke McMahon • featured scientist Takeshi Saito
illustrated by Janet Pagliuca

HEART SURGERY
by Sean Taitt • featured scientist Dr. Ram Kumar Subramanyan
illustrated by Janet Pagliuca

MARINE BIOLOGY
by Maria Madrigal Orozco • featured scientist Charnelle Wickliff
illustrated by Michelle Laurentia Agatha

POLAR SCIENCE
by Jocelyn Argueta
featured scientists Dr. Dieuwertje "DJ" Kast and Jocelyn Argueta
illustrated by Janet Pagliuca

VIRTUAL REALITY
by CaTameron Bobino • featured scientist Sharon Mozgai
illustrated by Janet Pagliuca

Photo Credits

Cover iStock.com/skynesher **8** KenStock/Pixabay **11** photo courtesy of Michelle Agatha **12** photo courtesy of Kris Durán **16–17** Jim Lane / Alamy Stock Photo **19** martin berry / Alamy Stock Photo **22–23** NewAfrica/Depositphotos.com **24** Jocelyn Castro; photo courtesy of Jasmin Sanchez **26–27** photo courtesy of Jasmin Sanchez; Jason Dominguez; © Pixago | Dreamstime.com; iStock.com/FatCamera **28–29** Jason Dominguez; iStock.com/aquaArts studio; © Bhommel | Dreamstime.com; © Najmi Arif Norkaman | Dreamstime.com; iStock.com/Antonio_Diaz **30–31** grejak/Depositphotos.com **32–33** iStock.com/Bojani **34–35** © Patrick Daxenbichler | Dreamstime.com **36–37** iStock.com/IvanJekic; photo courtesy of Jasmin Sanchez; photo courtesy of Michelle Agatha **40** photo by Allan Mas from Pexels